Prospects for Shale Gas Development in Asia

EXAMINING POTENTIALS AND CHALLENGES IN CHINA AND INDIA

Authors
Jane Nakano
David Pumphrey
Robert Price Jr.
Molly A. Walton

August 2012

About CSIS—50th Anniversary Year

For 50 years, the Center for Strategic and International Studies (CSIS) has developed practical solutions to the world's greatest challenges. As we celebrate this milestone, CSIS scholars continue to provide strategic insights and bipartisan policy solutions to help decisionmakers chart a course toward a better world.

CSIS is a bipartisan, nonprofit organization headquartered in Washington, D.C. The Center's 220 full-time staff and large network of affiliated scholars conduct research and analysis and develop policy initiatives that look into the future and anticipate change.

Since 1962, CSIS has been dedicated to finding ways to sustain American prominence and prosperity as a force for good in the world. After 50 years, CSIS has become one of the world's preeminent international policy institutions focused on defense and security; regional stability; and transnational challenges ranging from energy and climate to global development and economic integration.

Former U.S. senator Sam Nunn has chaired the CSIS Board of Trustees since 1999. John J. Hamre became the Center's president and chief executive officer in 2000. CSIS was founded by David M. Abshire and Admiral Arleigh Burke.

CSIS does not take specific policy positions; accordingly, all views expressed herein should be understood to be solely those of the author(s).

Cover photo: Drilling rig and worker, © iStockphoto.com/shuda/Huiping Zhu.

ISBN 978-0-89206-742-8

Center for Strategic and International Studies
1800 K Street, NW, Washington, DC 20006
Tel: (202) 887-0200
Fax: (202) 775-3199
Web: www.csis.org

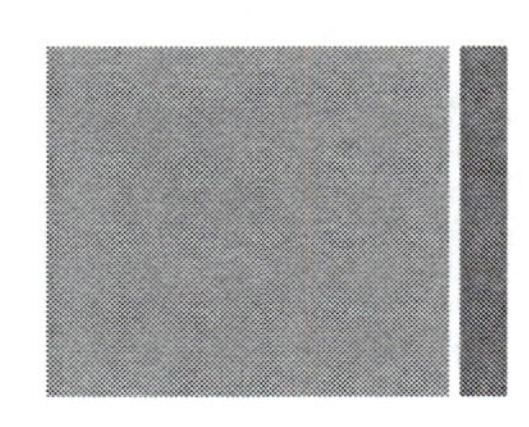

CONTENTS

ACKNOWLEDGMENTS

The authors gratefully acknowledge the valuable assistance of their CSIS colleagues throughout the production of this report: Frank Verrastro, Sarah Ladislaw, Lisa Hyland, and our studious interns—Clare Richardson-Barlow, Youngji Jo, Lin Shi, and Mallory Lee Wong. Also, our gratitude goes to Bo Kong, Dagmar Graczyk, Don Juckett, Kang Wu, Ksenia Kushkina, and Mark Stern for sharing their valuable insights into Chinese and Indian energy sectors.

Any errors or omissions are the responsibility of the authors.

EXECUTIVE SUMMARY

China and India are two of the world's fastest-growing economies, and their economic growth drives a strong demand for energy and natural resources. Between now and 2035, global energy consumption is forecast to grow by 50 percent, and China and India together will account for more than half of this global growth. The scale of their energy consumption affects global supply and demand and, inherently, the price levels of various energy commodities—including natural gas—in the global marketplace.

The development of unconventional gas resources, especially shale gas, in China and India warrants close observations because of a host of potential economic and energy security benefits successful development may bring for the two growing economies. An April 2011 assessment of international shale gas resources by the U.S. Energy Information Administration (EIA) cited technically recoverable shale gas resources (*not* reserves) in China at 1,275 trillion cubic feet (tcf) and in India at 63 tcf, compared with 1,250 tcf for the United States and Canada combined. China and India have already begun exploring their substantial indigenous shale gas resources, but the question of how well they can replicate the U.S. experience—and over what time period—still looms large. The geological characteristics of shale deposits can vary widely, affecting the potential production profiles.

However, the volume of geological resources is only one side of the coin. A host of "above-ground" conditions are essential in fostering the successful development of these resources. In the United States, the so-called shale gas revolution resulted from the confluence of factors, including access to shale gas resources on private lands, economically attractive natural gas prices (2007–2008) that spurred investment interest, innovative operational and technological step changes that combined hydraulic fracturing (fracking) with extended-reach lateral wells, an evolving understanding of how shale formations react to stimulation, and availability of infrastructure to process and transport the gas. The results have thus far been nothing short of extraordinary, and shale development, which accounted for a negligible amount of U.S. natural gas production less than a decade ago, now makes up almost 50 percent of domestic output.

Both policy pronouncements and emerging investments into North American shale basins suggest that Chinese and Indian interests in exploring the potential of their unconventional gas resources, especially shale gas, are real. However, much still needs to be known about the commerciality of their resources. China and India would benefit from the availability of reliable data and processing capability. Additionally, both countries have yet to fully formulate—although China appears ahead of India—policy frameworks concerning regulatory and physical infrastructure, pricing mechanisms, and environmental and resource management as well as issues associated with societal challenges that may accompany a large-scale development of their unconventional gas resources.

Also, the availability of technical expertise to manage and deploy advanced exploration and production technologies would be another factor in determining the pace and scale of shale gas development in China and India. All of these factors suggest that the pace of development of China's and India's shale gas resources could be significantly slower than the North American experience.

1 SHALE GAS IN CHINA

As one of the world's largest and fastest-growing energy consumers, China is focused on securing new and diverse supplies of energy to maintain a healthy rate of economic growth and societal development. Natural gas has recently become a key area of focus for domestic resource exploitation and increased areas of trade (i.e., imports). China has significant shale gas potential that it wishes to develop, although significant investment, infrastructure, and policy and market barriers must be resolved to realize the full potential.

Natural Gas—A Burgeoning Fuel Choice

Throughout the 1990s, Chinese leadership largely dismissed natural gas, believing it to be too expensive to compete with domestic coal resources. In the early 2000s, driven mainly by growing environmental concerns, China began pushing gas development: domestic production, expansion of the gas pipeline network, and imports of both pipeline gas and liquefied natural gas (LNG). Demand continues to outstrip supply, making China a net importer, but domestic production has been on the rise. China increased domestic production from 27.2 billion cubic meters (bcm) (960.56 billion cubic feet [bcf]) in 2000[1] (approximately 106 percent of domestic consumption) to 94.5 bcm (3.34 trillion cubic feet [tcf]) in 2010 (approximately 89 percent of domestic consumption). It has constructed the 4,200-km West-to-East pipeline to carry gas from Xinjiang Province in the west to Shanghai, opened the TransAsian pipeline in 2009 to bring gas from Turkmenistan to China, and brought four LNG receiving terminals online, allowing LNG to meet some 10 percent of Chinese gas demand. A few more LNG receiving terminals are under construction.

The main method of encouraging expanded production and use of natural gas is through government-set targets and mandates, which state-run companies, with some participation from private companies, then follow up with investments and projects. So far, China has increased the share of natural gas in total energy requirements from 2 percent to 4 percent, but it plans to reach 8 percent by 2015 and 10 percent by 2020. China's Ministry of Land and Resources' (MLR) hydrocarbon research arm is forecasting shale gas production in China to reach 6.5 bcm (229.55 bcf) a year by 2015, which is equivalent to 6.4 percent of China's total gas production today. China's 2020 production target—issued jointly by the National Development and Reform Commission (NDRC), Ministry of Finance, MLR, and National Energy Administration—at 60–100 bcm (2.12–3.53 tcf) per year[2] would be equivalent to the entire volume of natural gas the country produces today.

1. BP, *Statistical Review of World Energy 2010,* London.

2. China Capital Stock Daily, "Shale gas will have market-based pricing and all kinds of incentives could be higher than traditional natural gas" [Ye Yan Qi Jiang Shi Xing Shi Chang Ding Jia. Ge Xiang Bu Tie You Wang Gao Yu Mei Ceng Qi], March 19, 2012, http://business.sohu.com/20120319/n338141282.shtml.

Estimates of China's longer-term natural gas production and consumption vary according to institution. BP's long-term outlook released in January 2011 posits natural gas demand growing in China at 7.6 percent per year over the next two decades, resulting in gas use of 444 bcm (15.6 tcf) in 2030, about 10 percent lower than current gas demand in the European Union. This still would represent only 9 percent of total energy requirements.[3] Other forecasts are less optimistic. ExxonMobil's long-term outlook puts China's 2030 natural gas consumption at only 281 bcm (9.92 tcf).[4] In the U.S. Energy Information Administration (EIA) forecasts, gas holds a 5 percent share of total energy in 2020 and edges up to only 6 percent by 2035.[5] Predictions from Chinese institutions are much more sanguine, however. According to Jiping Zhou, vice president of the China National Petroleum Corporation (CNPC), the demand for natural gas in China could reach 230 bcm (8.12 tcf) in 2015, 350 bcm (12.36 tcf) in 2020, and 500 bcm (17.66 tcf) in 2030.[6]

Unconventional Gas Resources

Until recently, unconventional gas in China has been focused primarily on coalbed methane (CBM). Coalbed methane is natural gas found in association with coal resources. China's vast coal resources and coal mining activities made CBM a natural place to start in the field of unconventional gas exploitation. China holds the third-largest CBM resource base in terms of geological volume, following Russia and Canada. There are nine major CBM basins in China: Odors, Qinshui, Junggar, Diandongqianxi, Erlian, Tuha, Tarim, Tianshan, and Hailaer. The CBM resources from these basins amount to 30.9 trillion cubic meters (tcm) (1,091 tcf)—roughly 84 percent of the total CBM resources of China.[7] According to the 12th Five-Year Plan (2011–2015), the production target for CBM is 21.5 bcm by 2015,[8] up from 9.1 bcm in 2010.[9] China's National Energy Administration (NEA) states that China plans on investing 116.6 billion yuan ($18.4 billion) in CBM production over the next four years and establishing 13 pipelines with a capacity of 12 bcm per year.[10]

Chinese commercial interests in CBM exploration and production seem to be on rise. For example, PetroChina CBM Co. plans on spending more than $1.5 billion in the next three years to increase its CBM production capacity by 4.5 bcm (158.9 bcf), which is equivalent to nearly double

---

3. BP, *BP Energy Outlook 2030,* London, January 2011, http://www.bp.com/energyoutlook2030.

4. ExxonMobil, *The Outlook for Energy: A View to 2030*, Houston, December 2010, http://www.exxonmobil.com/Corporate/files/news_pub_eo_2010.pdf.

5. U.S. Energy Information Administration, *International Energy Outlook 2010,* Report DOE/EIA-9484 (2010), Washington, D.C., July 2010, http://www.zerauto.nl/cp/uploads/bronnen/EIA%20%20outlook%20 2010_1298298031.pdf.

6. China National Petroleum Corporation, "China's Natural Gas Demand and CNPC's Natural Gas Business Strategy," November 30, 2010, http://www.cnpc.com.cn/en/press/speeches/China_s_Natural_Gas_Demand_and_CNPC_s_Natural_Gas_Business_Strategy_.htm.

7. D.K. Luo et al., "Economic Evaluation Based Policy Analysis for Coalbed Methane Industry in China," *Energy*, November 18, 2011, http://cat.inist.fr/?aModele=afficheN&cpsidt=23743531.

8. Zhou Yan, "Rules to govern foreign participation in CBM," *China Daily*, November 4, 2011, http://www.chinadaily.com.cn/business/2011-11/04/content_14035589.htm.

9. Ibid.

10. Xinhua, "China's coal-bed methane output to reach 30bln cubic meters in 2015," December 31, 2011, http://news.xinhuanet.com/english/china/2011-12/31/c_131337443.htm.

China's entire capacity today.[11] The plan entails projects in the Xinjiang and Inner Mongolia autonomous regions as well as Shanxi Province. Sinopec Group announced plans in 2010 to boost its unconventional gas production, including CBM and shale gas, to 2.5 bcm (88.29 bcf) annually by 2015.[12]

Foreign entities have long been involved in the CBM industry in China and, according to China Daily, foreign funding has accounted for nearly 70 percent of the CBM exploration there.[13] According to statements by PetroChina CBM president Jie Mingshun, China's CBM industry continues to attract foreign players: Shell is interested in cooperating with PetroChina to develop a CBM site in Erdos, while BP has already agreed to cooperate with PetroChina in CBM development in the Tuha Basin in Xinjiang.[14] Also, since 2002, Far East Energy Corp. of Texas has been working closely with China United Coalbed Methane Co. Meanwhile, Jie points to the use of horizontal drilling technology in a multidivided or feather-like spread from one or more wells as a primary bottleneck to CBM development, and underlines the desire for breakthroughs in key technology.[15] Furthermore, Jie hopes that only one-third of the target level will be the result of foreign cooperation.[16]

Uncertain Shale Gas Resource Estimates and Potentially Promising Areas

Except for coalbed methane, which is estimated to be 1,306.64 tcf (37 tcm), there is no consensus on the amount of unconventional gas resources in China. The estimates on the amount of shale gas resources vary greatly:

Institution	Shale Resource Estimate
U.S. Energy Information Administration	1,274.85 tcf
International Energy Agency	918.18 tcf
China Ministry of Land and Resources	886 tcf
China National Petroleum Corporation	1,084 tcf

Many Chinese national oil companies can offer insights into the shale gas resource potential in China. According to CNPC, the geological conditions of the Sichuan Basin in central China, the Ordos Basin in western China, the Middle and Lower Yangtze Platform regions in southern China, and the North China Basin present the best potential for shale gas resources. The Junggar, Songliao, and Turpan-Kumal Basins are good prospects, while potential in the Qaidam Basin and

11. Chen Jialu, "Energy Company Taps Methane as Energy Needs Surge," *China Daily,* October 26, 2010, http://www.chinadaily.com.cn/usa/2010-10/26/content_11458884.htm.

12. John Duce and Chua Baizhen, "Chevron, Sinopec in Talks to Develop Shale Gas as China Seeks to Boost Use," Bloomberg News, September 17, 2010, http://www.bloomberg.com/news/2010-09-17/chevron-says-in-discussions-with-sinopec-group-on-shale-gas-cooperation.html.

13. Chen Jialu, "Energy Company Taps Methane as Energy Needs Surge."

14. Ibid.

15. Ibid.

16. Ibid.

Liaohe Province of the Bohai Bay Basin is poor.[17] CNPC has compared Chinese basins with U.S. analogs to obtain an estimate of shale gas resources. The Yangtze and North China were compared with the Appalachia cratonic basin, the Western China Continental foreland basin was compared with the San Juan Basin, and the Western China Continental depressed basin was compared with the Michigan basin. Based on this analog analysis, CNPC has estimated the shale gas resources of China's major basins and regions at 21.5–45.0 tcm (759–1,589 tcf), with a median estimate of 30.7 tcm (1,084 tcf).[18] Shale gas resources are located primarily in the south (46.8 percent) and northwest (43.0 percent). In terms of geology, the Paleozoic period dominates (66.7 percent), followed by the Mesozoic (26.7 percent) and Cenozoic (6.6).[19]

Map 1. Major Shale Gas Basins and Pipeline System of China

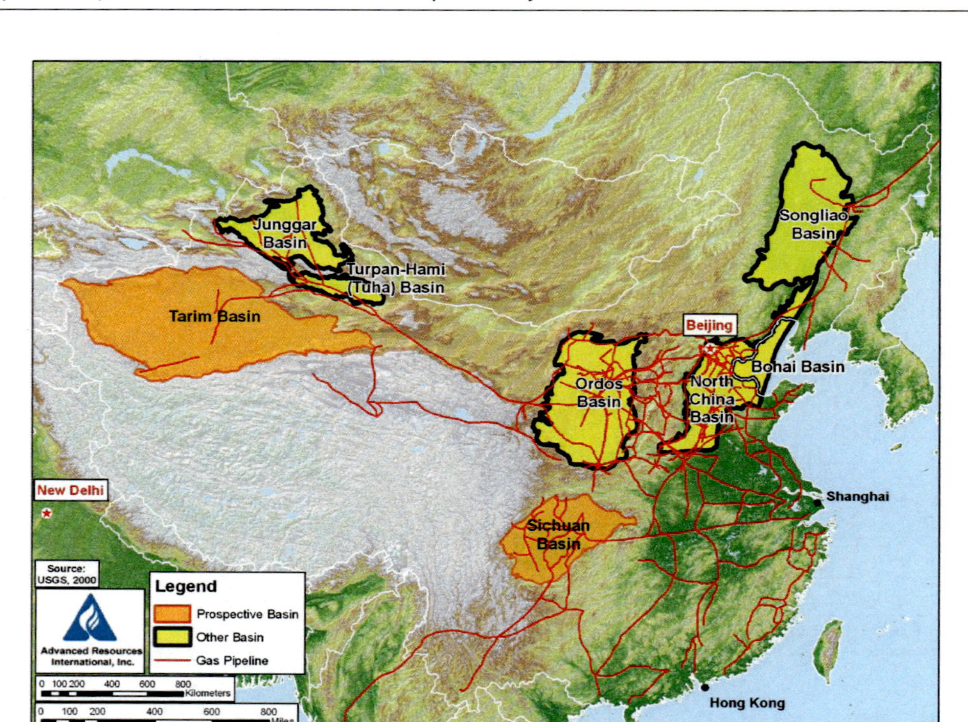

Source: U.S. Energy Information Administration (EIA), "Figure XI-1. Major Shale Gas Basins and Pipeline System of China," in *The World Shale Gas Resources: An Initial Assessment of 14 Regions outside the United States*, Washington, D.C., April 2011, p. XI-1.

17. Liu Honglin, "Progress of Exploration and Development of Shale Gas in China," presentation for China National Petroleum Corporation at the 10th U.S.-China Oil & Gas Industry Forum, Fort Worth, Texas, September 15, 2010.

18. Ibid.

19. Zhang Jinchuan, Fan Tailiang, and Yu Bingsong, "Resource Potential and Development Foundation of Shale Gas in China," presentation for the China University of Geosciences, Beijing, at the 10th U.S.-China Oil & Gas Industry Forum, Fort Worth, Texas, September 15, 2010.

The lack of comprehensive data on nationwide shale resources is fully acknowledged in China's *Shale Gas Development Plan for 2011–2015*, released in March 2012.[20] The Chinese energy planners recognize technological and financial challenges associated with assessing large-scale resources and then drawing development plans accordingly.[21] Under the 12th Five-Year Plan, China plans on completing the initial assessment for shale gas and confirming the current reserve estimates.[22] Additionally, the *Shale Gas Development Plan for 2011–2015* stipulates the establishment of a "national shale gas resource potential evaluation and advantageous region" sub-project under the national oil and gas strategic region special project. Under this initiative, the country is divided into five large regions—the upper Yangzte and Yunnan-Guizhou-Guangxi region, the middle to lower Yangtze and southeast region, the northern and northeastern region, the northwest region, and Tibet—for resource assessment and evaluation.[23] During the 13th Five-Year Plan (2016–2020), China plans on scaling up the shale gas development and exploration of 19 regions while developing new shale gas resources in regions like Hubei-Hunan provinces, Jiangsu-Zhejiang-Anhui provinces, Ordos, the southern part of the northern region, Songliao region, Zhunger Basin, Turfan Qomul region, Tarim Basin, and Bohai Sea Gulf.[24] The Chinese government also notes that gas-rich regions in China are highly populated, rendering exploration work there complex.[25] China is still very much in the process of evaluating its shale resource base. Despite the written objectives of the Chinese government, only time and future development will tell what potential resources exist and how much is ultimately recoverable over a given time period.

Will Shale Gas Production Displace the Need for LNG or Pipeline Imports?

Although China has only begun to explore its shale gas reserves, a number of global industry analysts already predict that China's shale gas production has the potential to cut into its demand for imported LNG no later than 2020. In September 2010, Australia's Macquarie Bank cited both Chinese shale gas development and Russian gas pipeline projects as a risk to Australian exporters of LNG to China, while Germany's Deutsche Bank warned that Australian coal seam gas-based LNG export schemes could be threatened by shale gas output in China.[26] UK industry consultants Wood Mackenzie believe China will need much less LNG beyond 2020, compared to the current decade, and that unconventional gas in China—shale gas, coal gasification, and CBM—will supply more than 12 bcf per day by 2030.[27] Given the long lead-times to develop these unconventional resources, especially shale gas, Wood Mackenzie still sees a sharp rise in China's LNG imports to 46 million tons (2.2 tcf) in 2020, compared to their earlier forecast of 31 million tons (1.61 tcf),

20. National Development and Reform Commission, Ministry of Finance, Ministry of Land and Resources, and National Energy Administration, *Shale Gas Development Plan 2011–2015* [Ye yan qi fa zhan gui hua (2011–2015 nian)], March 13, 2012.

21. Ibid., p. 4.

22. Ibid., p. 7.

23. Ibid.

24. Ibid., pp. 10–11.

25. Ibid., p. 3.

26. Matt Chambers, "LNG demand worries," *The Australian,* September 18, 2010, http://www.theaustralian.com.au/business/lng-demand-worries/story-e6frg8zx-1225925618741.

27. Carola Hoyos, "China gas growth to hit western groups," *Financial Times,* July 25, 2010, http://www.ft.com/intl/cms/s/0/7acd8eda-9820-11df-b218-00144feab49a.html#axzz1zTzh7GFR.

and to 54 million tons (2.8 tcf) in 2030.[28] China National Offshore Oil Co. (CNOOC) chairman Fu Chengyu said in 2009 that China planned to import 60 million tons (3.1 tcf) of LNG by 2020.[29] With four more LNG receiving terminals under construction and many more under consideration, not including expansions already planned at Shenzen, Fujian, and Shanghai, China will have the terminal capacity to reach its ambitious LNG import goals and CNOOC, PetroChina, and Sinopec already have entered into deals to supply some of the future terminals. The fact that some of these deals include upstream equity positions by the Chinese importers makes the agreements particularly robust. The Wood Mackenzie study also doubted China's need for any additional pipeline gas imports after 2020.[30]

Yet-to-Be Determined Policies and Regulations

China has yet to finalize shale gas production policy, as there has not yet been consensus on the best way to proceed. Many analysts suspect the Chinese policy on the development and use of shale gas will likely mirror that of CBM. In other words, the government probably will employ a combination of (1) import tax reduction or exemption for technology imports that are used for shale gas exploration; (2) exemption of prospecting and mining royalties; and (3) production subsidy—likely in the range of 3 to 5 cents per cubic meter (0.085 to 0.142 cents per cubic foot). Because the policy on CBM development will likely serve as a basis, or a reference at least, for formulation of shale gas development policy, following is an outline of policy measures that have been employed for CBM development:[31]

- ***Resource management policies.*** In order to address the separation of mineral rights and development rights (i.e., in China, a CBM company often does not have the coal mining right, and a coal company often does not have the CBM development right), the MLR issued in 2007 a regulatory amendment on strengthening the management of integrated prospecting and mining of coal and CBM resource. The resource management policies for shale gas thus far states that shale gas resources should be managed as an independent mineral,[32] and encourages competition and exploration of resource potentials by those that own oil, gas, and CBM rights.[33]
- ***Environmental policies.*** The Ministry of Environmental Protection promulgated the Emission Standard of CBM/CMM (coal-mine methane) in 2008. This standard covers new coal mines, surface drainage systems, and existing mines/systems. This standard prohibits CBM drainage systems from emitting CBM, and coal mine drainage systems from emitting gases with a methane concentration of 30 percent or higher (e.g., they must either use or flare the gas).
- ***Foreign cooperation policies.*** Exploration and development of CBM through foreign cooperation should comply with Chinese regulations concerning on-shore petroleum resources. The

28. UPI, "China's demand for LNG to soar," UPI.com, July 26, 2010, http://www.upi.com/Business_News/Energy-Resources/2010/07/26/Chinas-demand-for-LNG-to-soar/UPI-91961280166774/.

29. Xinhua (New China News Agency), "China to Import More LNG from Qatar," EnergyCurrent.com, March 9, 2009, http://news.xinhuanet.com/english/2009-03/08/content_10965858.htm.

30. Syed Rashid Husain, "Beijing may not require additional gas imports by 2020," ArabNews.com, September 11, 2010, http://www.arabnews.com/node/354969.

31. Luo et al., "Economic Evaluation Based Policy Analysis for Coalbed Methane Industry in China."

32. China, *Shale Gas Development Plan for 2011–2015*, March 2012, p. 2.

33. Ibid., p. 12.

cooperation must be based on contracts, and the current standard contract is production-sharing contract (PSC).

- ***Technology R&D policies.*** Under the National Key Technologies Research and Development Program of 1983, China has been supporting the development of CBM exploration and development technology. Furthermore, the government has been promoting the development of innovative oil and gas exploration and development theory and technology under the National Medium and Long-term Science and Technology Development Program (2006–2020).
- ***Value-added tax (VAT).*** For all CBM companies, VAT will be reimbursed after being levied. City maintenance and construction tax, education surcharge, and some other local taxes are levied at 10 percent. Import duties, import-related tax, and VAT will be exempted for CBM exploration and development operations, equipment, spare parts, and special tools.
- ***Corporate income tax.*** For the self-operating CBM companies in China, the corporate income tax has been levied at a reduced tax rate of 25 percent since 2008. The CBM companies cooperating with foreign companies are entitled to preferential corporate income tax policy whereby corporate income tax is exempted in the first two years from the profit-making year, and then levied with a 50 percent reduction in the ensuing three years.
- ***Resource tax.*** Since 2007, companies engaged in surface recovery of CBM are exempt from resource tax.
- ***Subsidies.*** Per the Ministry of Finance's Executing Opinions on Subsidizing CBM/CMM Development and Utilization, enterprises engaged in CBM/CMM extraction within China are entitled to a financial subsidy of 0.2 yuan per cubic meter (0.029 USD per cubic meter) from the central government if the gas is used on site or marketed for residential use or as a chemical feedstock. According to the CBM development program published by NEA in late December 2011, the government plans to increase the production subsidy to 0.4 to 0.5 yuan per cubic meter.

Who Are the Key Players and Stakeholders?

Key government players in the Chinese development of shale gas are the Ministry of Land and Resources (MLR), National Energy Administration (NEA), Ministry of Commerce, Ministry of Finance, Ministry of Environmental Protection, and Ministry of Water Resources. Of these, MLR is the most active as it sees revenue-making opportunities through licensing. MLR collects royalties and fees from mining prospects. Meanwhile, NEA is supportive of unconventional gas exploration as means of addressing China's greenhouse gas emission challenges, but also to help reduce the country's dependence on fossil fuel imports.

While the government contemplates the best mix of policy and measures, Chinese national oil companies (NOCs) are starting to lock up land and resources to prepare for when these policies are enacted. Chinese NOCs dominate the shale gas production areas currently being made available for exploration—though it is expected that international companies will be involved in several different ways. In June 2011, MLR issued its first shale gas exploration tender with an offer of four blocks—mostly in the southwestern Chongqing municipality and Guizhou province—to a group of Chinese companies: CNPC subsidiary PetroChina, Sinopec, CNOOC, Shaanxi Yanchang Petroleum Group, China United Coal Bed Methane, and Henan Provincial Coal Seam Gas Development and Utilization Co.[34] According to the International Energy Agency, participants in this licensing round must commit to both a minimum investment and a minimum number of wells to be drilled and hydraulically fractured so as to maximize exploration within the offered acreage and to assist Chinese companies to acquire fracturing knowledge.[35] Although the government initially offered four blocks, two were canceled as they did not receive enough bids.[36] In July 2011, Sinopec Corp. and a provincial gas company won the exploration rights to the Nanchuan block while Henan Province Coal Seam Gas Development and Utilization Co. won the Xiushan block.[37] According to MLR, Sinopec plans to invest 591 million yuan ($91 million) on exploration at the block, and Henan's Coal Seam Gas plans to spend 248 million yuan.[38]

While current lease sales have excluded international energy companies, many companies are positioning themselves to participate in unconventional gas activities in China through various channels. International energy companies also seek a more level playing field across the natural gas sector and hope to convince Chinese policymakers of the value and experience they bring to the table as part of their overall effort to create a more developed, efficient, and well-functioning natural gas market in China.[39]

Emerging Shale Gas Industry

The world has seen a rise in shale gas-related investment by Chinese companies in recent years. Much of this investment has centered around strategic investments in oil and other minerals via direct purchase of assets, bidding on the right to produce the resource, or buying an equity stake in a company with ownership or production rights. In 2010, Chinese NOCs expanded their interest to the unconventional gas resources in North America and Australia. For example, in November 2010, CNOOC Ltd. completed a $1.08 billion purchase of a one-third interest in South Texas Eagle Ford Shale Basin assets—owned by the U.S. company Chesapeake Energy Corp.—and in January 2011, a $570 million purchase of one-third of the Niobrara shale project in Colorado and Wyoming—also owned by Chesapeake Energy Corp. Most recently, in January 2012, Sinopec agreed to buy a one-third stake in five Devon Energy Corp. exploratory oil projects in the U.S. for $900 million.[40]

Some analysts speculate that Chinese NOCs are particularly interested in gaining access to technologies and experience in the context of the ventures in order to expedite the learning process that must come along with unconventional shale gas production. International oil and gas companies are developing strategies to protect technologies or experience to preserve competi-

34. Chen Aizhu et al., "China Kicks Off First Shale Gas Tender," Reuters, June 28, 2011, http://www.reuters.com/article/2011/06/28/shalegas-idUSL3E7HS0D620110628.

35. International Energy Agency, "World Energy Outlook 2011—Special Report: Are We Entering a Golden Age of Gas?" 2011, p. 57.

36. Shanghai Daily, "2 Firms Win Exploration Rights to Sichuan Shale Gas," July 8, 2011, http://www.china.org.cn/business/2011-07/08/content_22950543.htm.

37. Ibid.

38. Ibid.

39. Energy Working Group, "Position Paper," European Union Chamber of Commerce in China, 2010/2011.

40. Jim Polson and Benjamin Haas, "Sinopec Group to Buy Stakes in Devon Energy Oil Projects," Bloomberg, January 3, 2012, http://www.bloomberg.com/news/2012-01-03/sinopec-agrees-to-pay-900-million-for-stakes-in-five-devon-shale-projects.html.

tive practices where necessary and possible, while still taking advantage of the sizable commercial interest put forth by Chinese and other foreign companies.

U.S.-China Public Sector Engagement

Bilateral shale gas cooperation is gaining strong momentum in the U.S. and Chinese governments as well as from the business communities of both countries. In November 2009, U.S. president Obama and Chinese president Hu announced the launch of the U.S.-China Shale Gas Resource Initiative that covers the resource assessment, utilizing experience gained in the United States; technical cooperation to support accelerated development of shale gas resources in China; and investment promotion, study tours, and workshops. The U.S.-China Shale Gas Resource Initiative is part of the Unconventional Gas Technical Engagement Program, led by the U.S. Department of State and launched in April 2010, in order to assist countries seeking to utilize their unconventional natural gas resources—including shale gas—to identify and develop them safely and economically.

The two countries have also been engaged in information exchange at the working level. Shale gas has been among the topics of discussions at the U.S.-China Oil and Gas Industry Forum (OGIF) since its ninth meeting in Qingdao, China, in September 2009, where U.S. and Chinese government and industry participants gave presentations on resource and technical issues.[41] The Chinese delegation specified their need for technical expertise to evaluate and screen formations, and technologies to drill and fracture for the development of shale gas. The tenth OGIF in Fort Worth, Texas, in September 2010, furthered the exchange on resource and technical issues, and included a tour of drilling, production, and water-management sites at Chesapeake Energy's Barnett Shale. Also, the 11th OGIF in Chengdu, China, in September 2011, featured presentations on U.S. case studies and a visit to the Weiyuan shale gas field.

Additional technical exchanges have also been led by the U.S. Department of Energy (DOE). For example, in April 2010, DOE organized a three-day workshop in Beijing, where DOE experts briefed the Chinese on shale gas development experiences in the United States, including horizontal well design, hydraulic fracture design and evaluation, and water management.

Furthermore, at the U.S.-China Strategic and Economic Dialogue, in Beijing in May 2012, the United States and China agreed on strengthening cooperation and discussions on regulatory and environmental frameworks that would encourage responsible production of shale gas.

41. US-China Oil and Gas Industry Forum, "Ninth Forum Agenda," September 27–29, 2009, http://www.uschinaogf.org/Forum9/Agenda9_eng.pdf.

2 SHALE GAS IN INDIA

Like many other places in the world with shale gas potential, India recognizes the strategic importance of developing its shale gas resources. Shale gas could help meet the rapidly growing needs of a large and developing population while mitigating the need to increase imported natural gas via liquefied natural gas (LNG) or pipeline. India has significant governance, market, and industry hurdles to overcome before shale gas production can make a significant contribution to India's energy mix. Most analysts believe India is farther behind than China in almost all aspects of creating the right commercial frameworks for developing its shale gas potential.

Natural Gas—Visions and Priorities in Flux

While India's natural gas production has been on a steady rise, demand has been chronically outpacing supply and the country has been a net importer of natural gas since 2004. In 2009, India produced 39.3 bcm (1.39 tcf) and imported 12.6 bcm (0.44 tcf) as LNG—comprising 7 percent of the overall energy mix in India,[42] but price controls and government rationing and allocation of natural gas mean that conservatively some 30–35 bcm (1.06–1.24 tcf) of demand went unmet.[43] In 2010, India's net natural gas imports reached an estimated 429 bcf.[44] As a consequence, India has sought imports of natural gas to meet domestic demand while lowering the overall environmental impact of energy use. Like China, India relies heavily upon coal for energy, especially electric power production. The rush to find gas imports resulted in proposals for more than a dozen LNG terminals and examination of various pipeline import options.

So far only two LNG receiving terminals are in operation, both in the state of Gujarat. Shell/Total's Hazira terminal, on stream since April 2005, was expanded in 2008 and now has a capacity of 488 bcf per year.[45] Petronet LNG's 10 million tons per year (MMT/y) (487.09 bcf/y) terminal at Dahej came online in 2004, with an expansion completed in 2009.[46] Petronet is currently working on a second jetty at Dahej, required for risk mitigation, but also raising capacity another 2.5 MMT/y by allowing higher-capacity Q-Max and Q-Flex LNG ships to berth there. The 5 MMT/y Ratnagiri terminal at Dahbol, owned jointly by the National Thermal Power Co. and Gas Authority India, Ltd. (GAIL), has been complete for years, but continues to struggle and is not operation-

42. BP, *Statistical Review of World Energy 2010,* London, June 2010, http://www.bp.com/liveassets/bp_internet/globalbp/globalbp_uk_english/reports_and_publications/statistical_energy_review_2008/STAGING/local_assets/2010_downloads/statistical_review_of_world_energy_full_report_2010.pdf.

43. Vinayak Chatterjee, "Opinions & Analysis: India's gas story needs a solid base," *Business Standard,* February 23, 2011.

44. U.S. Energy Information Administration, "Country Analysis Brief: India," March 2012, http://205.254.135.7/countries/cab.cfm?fips=IN.

45. Ibid.

46. Petronet, "Dahej LNG Terminal," http://www.petronetlng.com/Dahej_LNG_Terminal.aspx.

al. Finally, Petronet's 2.5 MMT Kochi (Cochin) LNG terminal at Puthyvypeen Island in the state of Kerala is expected to come on stream in 2012, and Petronet's board has approved a later expansion of another 2.5 MMT. Still under consideration are India Oil Corp.'s 2.5 MMT Ennore plant, the Adani Group's 5 MMT Mundra LNG facility, and the 5 MMT Mangalore LNG terminal.

Map 2. Shale Gas Basins and Natural Gas Pipelines of India/Pakistan

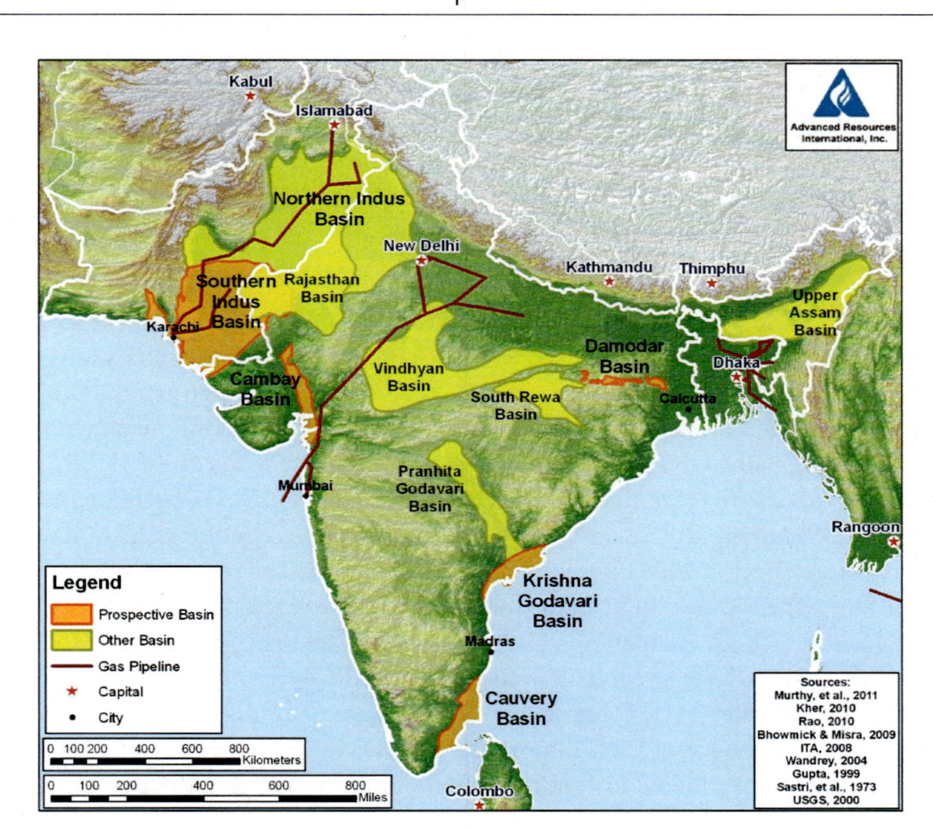

Source: U.S. Energy Information Administration (EIA), "Figure XII-1. Shale Gas Basins and Natural Gas Pipelines of India/Pakistan," in *The World Shale Gas Resources: An Initial Assessment of 14 Regions outside the United States*, Washington, D.C., April 2011, p. xii-1.

The country head for BP India, Sashi K. Mukundan, has indicated that as part of BP's recently announced $7.2 billion joint venture with Reliance Industries, BP will look at tapping its global liquefied gas resources to increase LNG imports into India, including possibly building a new LNG receiving terminal.[47] A senior Indian official said in February 2011 that India's LNG imports could hit 973.84 bcf (20 million tons) by 2017, more than twice the level of 2010.[48]

India also has considered several options for importing natural gas via pipeline, albeit the projects are of a questionable value from the supply security perspective and also face commer-

47. "BP-RIL deal to help develop domestic gas market: Sashi K Mukundan, PB, country head," *Economic Times,* February 25, 2011, http://articles.economictimes.indiatimes.com/2011-02-25/news/28634273_1_shale-gas-bp-ril-deal-country-head.

48. Reuters, "LNG traders flock to Singapore to tap China, India demand," February 28, 2011, http://www.reuters.com/article/2011/02/28/idUSL3E7DS04A20110228.

cial obstacles. These include the Turkmenistan-Afghanistan-Pakistan-India (TAPI) pipeline; the Iran-Pakistan-India (IPI) pipeline; and pipelines from Myanmar to India. The $7.6 billion TAPI pipeline would transport 33 bcm/y (1.7 tcf/y) along a 1,271-km stretch from the Dauletabad field in southeastern Turkmenistan to Multan, Pakistan, with a 640-km extension to India. Some question whether Turkmenistan has over-contracted its gas for exports and what this may mean for the TAPI. The proposed 2,600-km, $7 billion IPI pipeline is designed to carry up to 150 million cubic meters per day (mmcm/d) (5.3 bcf/d) from the South Pars fields in Iran to Gujarat, in India, with India potentially taking up to two-thirds. However, no project agreement has been put in place due to a mix of economic, political, and security issues associated with the proposed project. Lastly, the India-Myanmar pipeline project is another uncertainty with disagreements arising over whether the pipeline should go through Bangladesh. Also, the natural gas supply deal signed in 2009, between Myanmar and China, would source the supply from a field invested in by GAIL and Oil and Natural Gas Corporation (ONGC), further complicating the prospect for the India-Myanmar project. Perhaps the security question associated with these pipeline projects demonstrates the true value to the development of domestic shale gas resources to meet India's future gas demand.

The natural gas reserves discovered in the Krishna-Godavari Basin's D-6 block (KG-D6), in the Bay of Bengal, were the largest gas discovery of 2002. The discovery uplifted the prospect for domestic gas production in India and dampened the expected pace and volume of natural gas import. At one point, Reliance Industries, a majority owner of the block, expected the production level to rise as high as 4.2 bcf/d (120 mmcm/d) by roughly the 2012–2014 time frame. Additionally, the KG-D6 production largely accounted for the 25.6 percent increase in domestic Indian natural gas production from 1.17 tcf in 2008 to 1.47 tcf in 2009.[49] However, the reserve levels for some of its largest fields now appear much lower than anticipated. Gas production in the fourth quarter of 2011 averaged 40.49 mmcm/d, compared to 55.21 mmcm/d in the same period a year earlier, and the current output is said to be less than half of the peak that Reliance Industries had projected.[50] Considering that India's shale gas exploration has only just begun, production of shale gas appears unlikely to influence pipeline gas or LNG imports during the coming decade. Even after 2020, rising demand for natural gas, linked to continued vigorous economic growth, may mean that any domestic shale gas production will be taken up without altering the need for external supplies.

Shale Resource Potential

India is at the very early stage in its efforts to evaluate the nation's shale resource base. An April 2011 assessment of international shale gas resources by EIA cited technically recoverable shale gas resources (not reserves) in India at 63 tcf (1.8 tcm), compared with 1,275 tcf for China, 1,250 tcf for the United States and Canada combined, and 51 tcf for Pakistan.[51] In addition to the Barren Measure Shale at Icchapur, near Durgapur, West Bengal, ONGC also intends to explore shale

---

49. U.S. Energy Information Administration, "International Energy Statistics–India," 2009, http://www.eia.gov/countries/country-data.cfm?fips=IN.

50. "RIL's KG-D6 gas fields may hold lower reserves," *Times of India,* March 16, 2012, http://timesofindia.indiatimes.com/business/india-business/RILs-KG-D6-gas-fields-may-hold-lower-reserves-Niko/articleshow/12283098.cms.

51. U.S. Energy Information Administration, "World Shale Gas Resources: An Initial Assessment of 14 Regions Outside of the United States," April 2011, http://www.eia.gov/analysis/studies/worldshalegas/pdf/fullreport.pdf.

reserves in the Cambay, Kaveri-Godavari, Cauvery, Indo Gangatic, and Assam-Arakan Basins.[52] ONGC started with the Damodar Basin as it already had coalbed methane operations there, the shale is shallow, and water is plentiful at the location for eventual hydraulic fracturing of the prospect.[53] Overall, India is proceeding at a much slower pace than China.

Shale Gas Policymaking and Key Players

As was true of China, CBM also has a longer history of development in India than shale gas and can therefore serve as a potential guide for how the government might design shale gas policy. If policymaking for the CBM is an indication, key government players would involve the Ministry of Petroleum and Natural Gas (MoPNG), to set exploration and production policy, and the Directorate General of Hydrocarbons (DGH), to license and monitor exploration and production process, as well as prepare bid documents. The licensing process for shale gas will likely be coordinated through the DGH. Additionally, the Petroleum and Natural Gas Regulatory Board (PNGRB), created in 2006, promotes competitive markets for petroleum and petro products. The first round of exploration licensing, originally set for late 2011, was postponed due to the objection of the Ministry of Environment and Forests. The ministry objected to opening acreage for shale gas drilling and fracking until more detailed environmental studies were conducted and evaluated.

India has no comprehensive set of rules and regulations regarding the exploration and development of its domestic shale gas and oil resources. DGH is expected to draft such rules and regulations prior to the first shale asset auction, now pushed back to 2013 or beyond to allow for environmental studies. However, the Indian government will likely adopt the same process for shale gas development as it employed in the initiation of domestic CBM development a decade ago. Although the CBM development process has been slow, the fiscal regime was set up favorably for producers and for a commercial takeoff.

While India has traditionally used a production-sharing contract for oil and gas, it introduced a tax-and-royalty contract for CBM, based on the production level. A state-level government would receive a 10 percent royalty.[54] Also, data packages (e.g., basin dockets and block packages for each concession offered) were compiled by DGH with support from the Central Mines Planning and Design Institute (under the Ministry of Coal). Data rooms were set up to allow for preliminary review by companies; and interested companies were then able to purchase the complete data packages.[55] The bidding and award process for CBM has been transparent in that everyone can participate in the bidding, the criteria are known, and the bid evaluation is open.[56] Additionally, companies have the right to sell CBM on the open market.

Much has to be done before the licensing if shale gas development is to have a successful takeoff: establishing a regulatory framework, formulating fiscal policy, and completing data analyses, among others. The environment ministry already has claimed a role in shale gas development,

52. Sriparna Neogi, "Shale Gas: The Dark Knight in Shining Armor?," Trak.in, The India Business Blog, February 1, 2011, http://trak.in/tags/business/2011/02/01/shale-gas-environment-disaster/.

53. Editors, "India: Durgapur shale gas well under assessment," *Oil & Gas Journal,* February 2, 2011, http://www.pennenergy.com/index/petroleum/display/0221865790/articles/oil-gas-journal/exploration-development-2/area-drilling/20100/february-2011/india_-durgapur_shale.html.

54. CSIS Expert Roundtable, "Unconventional Gas: Developments in China and India," February 24, 2011.

55. Ibid.

56. Ibid.

forcing a delay in the auction of shale gas blocks until after 2013 while the MoPNG conducts a comprehensive study of water issues.[57]

A Nascent Shale Gas Industry

Some momentum may be starting for foreign companies interested in becoming involved in India's shale gas industry, as indicated by the agreement between ONGC and ConocoPhillips in March 2012, to undertake joint studies of shale gas opportunities in India and North America. However, Sclumberger—under contract to ONGC—appears to be the only company thus far with a shale gas activity in India. Schlumberger's Asian division drilled its first shale gas well in India in January 2011. The well was completed in the Damodar River Valley in eastern India. Schlumberger will drill additional wells as part of a seven-stage pilot project in West Bengal and Jamarkand. In June 2011, Oil India Ltd. (OIL) also hired Schlumberger Asia to assist OIL in Indian shale gas exploration. Schlumberger will carry out a feasibility study regarding OIL's shale gas reserves in Assam and Rajasthan.[58]

Schlumberger estimates for the Indian shale gas resources are 600 to 2,000 tcf,[59] potentially larger than shale gas resources in China. However, the shale gas exploration in India is less advanced than in China. Commercial production in India is likely 5 to 7 years away.[60] Proving actual, commercially recoverable shale gas reserves will require considerably more drilling.[61] Like China's state-owned oil and gas companies, India's state-owned and private hydrocarbon companies continually search for overseas investments to make up for India's relative paucity of oil and gas reserves. In opening the first day of the Indian Parliament's budget session in February 2011, Indian president Pratibha Devi Singh Patil noted the government's encouragement for state-run petroleum firms to acquire overseas assets. "Identification and exploitation of shale gas potential are being given priority," she told the Parliament.[62]

However, India's oil company foreign investments have not been as aggressive or widespread as those of the Chinese oil companies. In some cases, the companies' investments have encountered complicating circumstances or in other cases it appears the Indian companies were not as willing to pay a premium for the resources. For the most part it appears that India's oil companies are still evaluating where and whether to make investments at this time. Following are the examples of India's nascent investments in North American shale plays. In June 2010, Reliance acquired a 45 percent stake in Pioneer Natural Resources' Eagle Ford Shale gas assets for $1.3 billion.[63] A few months later, Reliance entered a joint venture with Houston-based Carrizo Oil & Gas Inc.,

---

57. Gireesh Chandra Prasad, "Shale gas blocks auctions put off by a yr," *Financial Express,* March 18, 2011, http://www.financialexpress.com/news/shale-gas-blocks-auctions-put-off-by-a-yr/764129/.

58. Ajay Modi, "Oil India hires Schumberger for shale gas foray," *Business Standard,* June 17, 2011.

59. Mitul Thakkar, "ONGC Drills Country's First Shale Gas Well," *Economic Times,* January 10, 2011.

60. Vinod Dar, "Emerging International Shale Gas: Poland and China Lead," *Right Side News,* January 14, 2011, http://community.nasdaq.com/News/2011-01/emerging-international-shale-gas-poland-and-china-lead.aspx?storyid=53312.

61. Eric Yep, "India Plans to Sign Shale Gas Pact with U.S.," *Wall Street Journal,* October 18, 2010, http://online.wsj.com/article/SB10001424052702304410504575559731818653568.html.

62. Abhrajit Gangopadhyay and Mukesh Jogota, "India President: Taming Inflation is Top Priority," *Wall Street Journal,* February 21, 2011, http://online.wsj.com/article/SB100014240527487044766045761576 42000501006.html.

63. Erika Kinetz, "Reliance pays $1.3B for Pioneer's Texas shale gas," *Victoria [Texas] Advocate,* June 24, 2010, http://m.victoriaadvocate.com/news/2010/jun/24/bc-as-india-reliance-pioneer/.

acquiring a 60 percent interest in Marcellus shale acreage that had been held under Carrizo's joint venture with another company.[64] Additionally, GAIL acquired a 20 percent stake in Carrizo Oil and Gas Inc.'s Eagle Shale Ford acreage for $95 million in September 2011.[65]

U.S.-India Public Sector Engagement

For the last several years India has been a strategic target for improved relations by the U.S. government. Beginning during the George W. Bush administration, India's geostrategic position and status as a large and emerging economy has put India on the diplomatic map. A major area of focus for strategic engagement with India is energy, including nuclear power, clean energy technologies, and now shale gas resource potential.

As part of the visit of President Barack Obama to India in November 2010, the U.S. Department of State and India's Ministry of Petroleum and Natural Gas entered into a Memorandum of Understanding on Shale Gas Reserves. The MOU contains a "Shale Gas Work Plan," which specifies assistance to be provided to MPNG's Directorate General of Hydrocarbons (DGH) by the U.S. Geological Survey to assess India's shale gas resource potential in a number of areas and basins; to train Indian nationals in shale gas resource assessment, including production potential; and to publish results of the technical studies and resource assessments.[66] The joint shale gas group is part of a larger effort announced by President Obama and Prime Minister Manmohan Singh, which includes a joint clean-energy research and development center in India, with each government contributing annual funding of $5 million over five years, with matching investment from the private sector.[67]

In addition to the highly visible political-level engagement, the U.S. Department of State has led interagency initiatives on shale gas, including briefings, roundtable discussions, and field visits in the United States for Indian delegations that address the technical, commercial, and policy issues that will influence development of shale gas resources. The U.S. Trade and Development Agency (USTDA) has sponsored training of Indian inspectors in offshore safety practices at the U.S. Minerals Management Service (MMS, now the Bureau of Ocean Energy Management and the Bureau of Safety and Environmental Enforcement), as well as MMS review of inspection practices at offshore natural gas sites in India. USTDA has also funded feasibility studies for Reliance and Essar in CBM, a feasibility study for a national natural gas pipeline network in India, training on natural gas regulation, and joint funding with the U.S. Environmental Protection Agency for establishment of a CBM clearinghouse at the Central Mine Planning and Design Institute Limited (CMPDIL) facilities in Ranchi, India (also supported by the Ministry of Coal, Ministry of Petroleum and Natural Gas/DGH).

---

64. "Reliance joins Carrizo in Marcellus shale venture," *Oil & Gas Journal,* August 5, 2010, http://www.ogj.com/articles/2010/08/reliance-joins-carrizo.html.

65. Rakesh Sharma, "GAIL India Acquires Stake in Carrizo's Shale Gas Assets for $95 Mln," *Wall Street Journal,* September 28, 2011, http://online.wsj.com/article/SB10001424052970204138204576599982561160822.html.

66. "Memorandum of Understanding between the Department of State, Government of the United States of America, and The (sic) Ministry of Petroleum and Natural Gas, Government of India, on Shale Gas Resources," signed in Washington, D.C., on November 1, 2010, by the Department of State, and in New Delhi, on November 6, 2010, by the Ministry of Petroleum and Natural Gas.

67. Reuters, "India, US ink deal for cooperation in shale gas resources," *Economic Times,* November 8, 2010, http://www.ausib.org/India,-US-ink-deal-for-cooperation-in-Shale-Gas-resources_1017_AUSIB-News.html.

3 CHALLENGES AND OPPORTUNITIES FOR ASIA

The shale gas sectors in China and India have seen an increased level of investment activities and interests as well as resource assessments in recent years. While the resources are necessary, they alone are not sufficient for a shale gas sector to take off commercially. The above-ground conditions significantly affect the commercial viability of shale gas development in these countries. Some of the key nongeological ingredients for successful development include technology and technical expertise, environmental and natural resource considerations, regulatory infrastructure, and physical infrastructure.

Role of Technology

Technological advancement has been shortening the timeframe for shale gas development since the time of the Barnett play, whose discovery dates to the late 1970s. Although it took Barnett roughly 30 years, the development timeframe for subsequent plays, ranging from Marcellus (from 2003) to Haynesville (from 2005), took approximately three years. Do China and India have sufficient technologies and technical expertise to move ahead on shale gas development? If not, what role does technology transfer play?

There is a notable difference between the state of the oil and gas service industry in China and that in India. For some time, the Chinese service industry has been dominated by indigenous service companies that were spun off from Chinese national oil companies, particularly CNPC and its regional subsidiaries. These companies already play an active role in CBM development in China. However, India does not have similarly developed indigenous service companies. Thus, Indians are more open to engaging Western service companies, as they already are with Schlumberger, Halliburton, and others. The range and breadth of service company expertise also affect the pricing of services and thus affect commerciality.

Furthermore, unlike the Chinese market, the Indian market has thus proven to be more open to foreign investments into its shale gas development. For example, the foreign participants likely will be permitted to hold a majority share in a joint venture arrangement or total equity ownership.

There appears consensus among technology holders and technical experts that the key to successful development of shale gas is the combination of both expertise and the application of reservoir-specific technologies. Many of the technologies being deployed for shale gas exploration and production in North America are commercially available. However, the technical expertise of managing and deploying these existing advanced technologies is the key to successful takeoff of the Chinese or Indian shale gas industry. For example, Chinese energy planners note that the core technologies for shale gas development are not fully understood or acquired by Chinese industry experts,[68] and that the Chinese shale gas development would require a large amount of research and technological breakthroughs.[69] Among the areas for technological breakthrough identified in China's *Shale Gas Development Plan for 2011–2015* are the development of "specialized technological service companies" as well as specific technologies like resource evaluation, and horizontal drilling and completion.[70]

Technical expertise is largely based on firsthand experiences. Therefore, one key challenge for the Chinese and Indians may be the gap between intellectual capacity—which both populations possess—and operational capability. Also arising from this observation is a question as to how quickly one can bring local labor force up to speed; the answer appears to come more in terms of years than in months. For example, a well design is a major determinant of whether a well may produce an optimal level of output, but the ability to design an optimal well comes primarily, if not solely, from years of experience.

The future landscape for shale gas development will likely look different as it is displayed in various parts of the world. It is highly likely that the North American shale industry will continue to use the same set of equipment for the immediate future, but the way in which the equipment is managed and deployed may become quite different.

Regulatory Infrastructure and Pricing

Besides the availability of technologies and presence of expertise, the regulatory environment greatly affects the prospect for successful takeoff in Asia. The presence of sound regulatory regime and pricing mechanisms are among key ingredients in allowing needed investments and facilitating efficient usage of the resources.

China

Two areas where the regulatory environment in China can be improved are the fiscal and pricing regimes. In contrast to the U.S. royalty-based system through which developers pay a percentage of project revenues and corporate taxes on profits, under China's production-sharing contracts (PSCs), international companies must share a large portion of their output with the government (or government-owned companies), in addition to paying corporate taxes on profits. In a high-risk venture like shale gas exploration and development, such a PSC system lowers the relative attractiveness of China.[71] Still, the potential size of the Chinese market has attracted the attention of such major international oil companies as Royal Dutch Shell, BP, ExxonMobil, Chevron, and ConocoPhillips. In March 2012, Shell became the first company to sign a formal production-sharing agreement with a Chinese entity.[72] The agreement with CNPC will provide Shell with a real stake in China's production of shale gas.

---

68. National Development and Reform Commission, Ministry of Finance, Ministry of Land and Resources, and National Energy Administration, *Shale Gas Development Plan 2011–2015*, March 13, 2012, p. 3.

69. Ibid., p. 4.

70. Ibid., p. 10.

71. Datamonitor, "Is China Ready to Explore Its Shale Gas Potential?" Datamonitor.com, January 31, 2011, http://www.datamonitor.com/store/News/is_china_ready_to_explore_its_shale_gas_potential?productid=996CF1DC-0B69-4616-B5EA-EE70CDA5B451.

72. Joel Kirkland, "Shell inks agreement to develop China's shale plays," E&E Publishing, March 22, 2012.

Also, the Chinese system of natural gas pricing is a patchwork of market and administered prices, varying between classes of consumers, wholesale versus retail, onshore versus offshore, and so forth. The current system of price controls for gas and other energy sources impedes natural gas achieving its full potential. The wide gap between domestic and international prices of natural gas has been forcing Chinese firms to import gas at a loss. Without significant price and regulatory reform, the more conservative projections for China's natural gas use in 2020 and 2030 are more likely.

In December 2011, the Chinese government instituted experimental reforms with a natural gas price mechanism in Guangdong province and the Guangxi region, where natural gas prices are no longer kept artificially lower than the market price levels. The experimental price mechanism will be linked to the market price of fuel oil and liquefied petroleum gas imported to Shanghai, which is a Chinese hub for gas trading and consumption. This reform affects the price of gas paid to pipeline owners by local gas utilities (commonly known as "city gate prices") in the two regions, where gas prices are already closer to international market prices due to their reliance on natural gas imports. It remains to be seen how quickly and widely this price reform may spread across China. Meanwhile, China also announced in December 2011 its decision to liberalize wholesale or well-head prices for unconventional gas resources, including shale gas, CBM, and coal gas.

Additionally, the Ministry of Land and Resources—not national oil companies—controls shale acreage. Reportedly, a friction between the ministry and local governments over land rights to shale blocks has been attributed as one cause for delaying the pace of acreage auction in China.[73] To the extent a shale revolution may occur in China, it may look quite different from how the "revolution" came about in the United States, where many independent companies initiated and drove the process.

India

The fiscal and pricing regimes are areas of concern for the shale gas potential in India, too. Some of the challenges are how to streamline the tax regime for gas exploration and production, and to clarify the fiscal structure for gas swapping.[74] Related to these issues is whether shale gas, like other gas supplies, will be subject to allocation by the central government to favored sectors such as fertilizer production. The current uncertainty concerning the allocation of fiscal revenues between the central and state governments has contributed to the slow pace of shale gas development by obfuscating potential incentives state governments may have in supporting shale gas exploration and development.[75]

Additionally, difficulties such as long delays and even violence involved in acquiring land in India are well known and could delay shale gas development. West Bengal, where ONGC has just completed its first shale gas test well, exploded in March 2007 when protesters against the taking

73. "China Shale Gas Sector Tough for Foreigners," *Oilgram News* 90, no. 80 (April 24, 2012), p. 2.

74. Vinayak Chatterjee, "India's Gas Story Needs a Solid Base," *Business Standard,* February 23, 2011, http://www.business-standard.com/india/news/vinayak-chatterjee-indias-gas-story-needssolid-base/426146/.

75. Shubi Arora, "Shale Gas: An essential part of India's plan for energy independence," *Bar and Bench,* November 17, 2010, http://www.barandbench.com/brief/3/1113/shale-gas0an-essential-part-of-indias-plan-for-energy-independence.

of agricultural land in Nandigram were fired on and more than a dozen killed. In 2008, Indian industrial mogul Ratan Tata was forced to relocate the factory to produce Tata Motor's breakthrough Nano automobile from Singur, in West Bengal, to Sanand in Gujarat State after violence and intimidation in West Bengal.

Besides the acquisition issue, a key question associated with the land may concern the access to shale gas resources. Shale gas reserves are located where CBM blocks are, but most CBM blocks are already leased. According to DGH, it has already awarded 71 percent of the total coal acreage in India for CBM in the first four rounds of licensing.[76] Most likely, shale gas concessions will be awarded under separate licenses from CBM or conventional gas. This is likely to sanction drilling of shale gas in shale seams, while methane under CBM licenses will be extracted only from coal seams.[77] As such, conflict resolution policies will have to be adopted. Meanwhile, many of the players in CBM also will have an interest in shale gas, so it may become possible to secure licenses for both within the same general acreage.[78]

Physical Infrastructure

The transport and delivery of energy and natural resources are generally an expensive endeavor. The delivery of natural gas, including shale gas, is no exception. Even with a strong reserve estimate, the scope and speed at which shale gas may join the mix of domestic energy supply are affected by whether a country has reliable physical infrastructure in place.

China

With all major gas transmission lines owned by the state and committed to current conventional pipelines, moving newly found shale gas to Chinese markets may require expanding existing gas pipelines and laying new pipes.[79]

MLR research center deputy director Zhang Dawei said that shale gas development would be restrained pending expansion of China's natural gas pipeline system.[80] Currently, CNPC essentially monopolizes pipeline construction and operations in China. It allows some CBM to be put into the West-East Pipeline. Whether CNPC would allow for shale gas transport by the third-party via their pipelines is a great unknown. The 4,200-km West-East Pipeline currently connects the Tarim and Ordos Basins to markets in the Shanghai area. The second West-East Pipeline was completed in June 2011, although several sub-lines remain to be completed later this year. A recently completed 1,700-km, 12 bcm/y gas pipeline carries Sichuan Province gas to Hubei, Anhui, Jiangxi, Jiangsu and Zhejiang Provinces, and Shanghai. China is constructing 14,400 miles (23,174 km) of new gas pipelines between 2009 and 2015 to build out the current 21,000-mile (33,796-km) network.[81]

---

76. CSIS Expert Roundtable, "Unconventional Gas: Developments in China and India," February 24, 2011.
77. Mark Stern, managing director, Mt Energy Associates.
78. Ibid.
79. Datamonitor, "Is China Ready to Explore Its Shale Gas Potential?"
80. Chen Aizhu, "China to auction 8 shale gas blocks in Q1," Reuters, January 20, 2011, http://www.reuters.com/article/2011/01/20/china-shalegas-idUSBJI00254020110120.
81. U.S. Energy Information Administration, "Country Analysis Brief: China," November 2010, http://www.eia.gov/countries/country-data.cfm?fips=CH&trk=p1.

Chinese energy planners recognize the infrastructure challenge and note that developing transportation and storage facilities would help to expand shale gas development. They further note that some parts of the shale gas rich regions have existing pipeline networks, but small-scale LNG and compressed natural gas (CNG) technologies may prove helpful in supporting the early stage of shale gas development and utilization in China.[82]

India

While ONGC has identified several prospective shale gas basins in India, India will have to add main transmission pipelines and tie-lines to bring any shale gas finds to market before commercial production can begin.

A 2006 policy change by India's Ministry of Petroleum and Natural Gas diluted GAIL's gas pipeline monopoly by allowing foreign investors, as well as private and public Indian companies, to hold 100 percent equity in gas pipeline projects. Nonetheless, GAIL's ownership and operation of the existing 4,100-mile gas trunk line network gives it continuing control. Approximately 80 percent of natural gas consumed in India is transported through this trunk pipeline. GAIL plans to add nearly 3,800 miles by 2014.[83] Meanwhile, Reliance now operates the 1,440-km East-West Pipeline, which brings its offshore production from the east to west coasts of India. GAIL and Reliance reached agreement on March 17, 2011, to an arrangement that will facilitate natural gas deliveries in Andhra Pradesh with gas swapped with LNG deliveries in Gujarat.[84]

Sanjay Kaula, member of the governing council of the University of Petroleum, Dehradun, observed, "India would be putting the cart before the horse by showing urgency for production of shale gas while the pipeline infrastructure remains dismally underdeveloped."[85] GAIL and Reliance have an alternative commercial view on this issue because it is necessary to confirm proven gas reserves before justifying the expense of new pipeline construction. Each of these companies has demonstrated opposition to proposals for certain competitive pipelines where the volumes of natural gas to support multiple lines have not been established.

Environmental/Resource Considerations

In the context of the United States, the environmental debate has centered on methane gas and chemical contamination of water, as well as disposal of drilling fluids. However, methane is more a regulatory question than a resource-management question. Therefore, in examining resource questions for China and India, this report limits its focus to water for the quantitative and qualitative considerations—already the major national issues for both countries.

---

82. National Development and Reform Commission, Ministry of Finance, Ministry of Land and Resources, and National Energy Administration, *Shale Gas Development Plan 2011–2015*, March 13, 2012, p. 5.

83. U.S. Energy Information Administration, "Country Analysis Brief: India," March 2012, http://www.eia.doe.gov/countries/cab.cfm?fips=IN.

84. "Gail, RIL swap gas to help state," *Deccan Chronicle,* March 18, 2011, http://www.deccanchronicle.com/node/15889/print.

85. Noor Mohammad, "Where are the pipelines?" *Indian Express,* February 21, 2011, http://www.indianexpress.com/news/where-are-the-pipelines/752516/.

China

Without even considering water-intensive shale gas production, a recent study sponsored by the International Finance Corp. found that by 2030, China's demand for water will outstrip current supply by 200 bcm, with agriculture accounting for more than half, industrial demand—led by thermal power generation—taking nearly a third, and the rest for domestic use.[86] Specifically, projections hold that eight out of ten river basins will experience water shortages by 2030, the largest gaps coming in the Hai Basin (an estimated 23 bcm) and the Yangtze Basin (an estimated gap of 70 bcm).[87] The eastern Huang (Yellow) and the Huai River basins will also experience water constraints (in terms of both quantity and quality of water). Many of the industries using heavy water, such as coal, reside in these basins.[88]

Water distribution within China is subject to spatial and temporal disparities as nearly 80 percent of the water resides in the south, while half of the population and much of the industry and agriculture are located in the north.[89] Southern China withdraws mostly surface water to meet demand while Northern China has relied on groundwater.[90] In fact, the per-capita level of available water varies per basin. Northern basins have an average of 780 cm/y while southern basins average 3,630 cm/y.[91] Thus, some of the most desirable shale plays (e.g., Tarim and Ordos) reside in river basins that are already experiencing acute water scarcity. This constraint presents a challenge for technology that relies on use of significant quantities of water.

China is aware of the very large water-management issues facing the production of other energy resources. Many coal reserves in Inner Mongolia, Xinjiang, and Shanxi are untapped, and the limited availability of water has prompted plans to transfer the water needed to mine, process, and consume these reserves. One proposal is to pipe water from the Bohai Sea to a desalination plant in Xilinhot, a coal-mining city in Eastern Mongolia.[92] Another proposal already underway is the South-to-North Water transfer project, which when completed in 2030 could transfer up to 22 km^3 of water via three major transfers from the Yangtze Basin to the Huai-Hai-Huang Basins.[93] However, the drought that has plagued central and eastern China since January 2011 has drastically lowered levels in the Yangtze River Basin, most notably in the Danjiangkou Reservoir on the Han River, a Yangtze River tributary that is a vital component of the central part of the South-to-North Water Transfer project, calling into question the long-term availability and reliability of water for multiple users, including the production of natural gas.

---

86. The 2030 Water Resources Group, *Charting Our Water Future: Economic Frameworks to Inform Decision-Making,* 2009, http://www.mckinsey.com/App_Media/Reports/Water/Charting_Our_Water_Future_Full_Report_001.pdf, p. 10.

87. Ibid., p. 60.

88. Ibid., p. 59.

89. Peter H. Gleick, "China and Water," in *The World's Water 2008–2009: The Biennial Report on Freshwater Resources,* ed. Peter H. Gleick (Washington, D.C.: Island Press, 2009), p. 85.

90. Food and Agriculture Organization of the United Nations, "Aquastat Country Profile," http://www.fao.org/nr/water/aquastat/countries_regions/index.stm.

91. Xie Jian et al., *Addressing China's Water Scarcity: Recommendations for Selected Water Resource Management Issues* (Washington, D.C.: World Bank, 2009), p. 11.

92. Keith Schneider, "Bohai Sea Pipeline Could Open China's Northern Coal Fields," Circle of Blue, April 5, 2011, http://www.circleofblue.org/waternews/2011/world/desalinating-the-bohai-sea-transcontinental-pipeline-could-open-chinas-northern-coal-fields/.

93. The 2030 Water Resources Group, *Charting Our Water Future,* p. 60.

China's leadership has long recognized the nation's water problems, drafting the National Water Law in 1988 as the legal framework for water management. Yet, water management in China is highly fragmented, and overlapping mandates and conflicts in jurisdiction and goals are common.[94] Water quality is regulated under the Ministry of Environmental Protection, while the Ministry of Water Resources is responsible for regulating water quantity; often information is not passed between the two. China's 11th Five-Year Plan (2006–2010) established water resource management as a core goal and priority. For example, the worst drought in China in 60 years already has led to emergency measures, such as cloud-seeding, and in its first policy document in 2011, the Chinese government declared water conservation a policy priority and pledged 4 trillion yuan in funding.[95] Also, China's Shale Gas Development Plan for 2011–2015 acknowledges the importance of environmental evaluation, especially concerning water management, in association with shale gas resource development.[96] However, China's inability to enforce its laws and regulations has hampered the effectiveness of past efforts.[97]

India

India faces a large gap between current supply of water and projected demand, amounting to a shortfall by 50 percent (754 bcm) in 2030. This gap is driven by a rapid increase in demand for water for agriculture, coupled with a limited supply infrastructure. Energy production competes with irrigation, drinking water, and industry for the water supply. With only 200 cm of water-storage capacity per person, India's accessible, reliable supply of water amounts to 29 percent of its total water resource.[98] According to the Ministry of Water Resources, the energy sector will require 3 times the current level of water supplies by 2025, and 26 times by 2050.

By 2030, water demand in India will reach twice the level expected in China, driven largely by agricultural needs. At nearly 1,500 bcm, Indian water needs in 2030 will far exceed the country's current annual supplies of some 740 bcm and will place severe strain on many of India's major river basins.[99] Given the great concerns about water availability and quality in parts of India, the government will need to include environmental rules and regulations as it promulgates the regime for shale gas development in India. These concerns already have led to a delay in India's first auction of shale gas leases.

94. Xie et al., "Addressing China's Water Scarcity," p. 30.
95. Krishnan Ananth, "In China, record drought brings focus on water security," *The Hindu,* February 12, 2011.
96. National Development and Reform Commission, Ministry of Finance, Ministry of Land and Resources, and National Energy Administration, *Shale Gas Development Plan 2011-2015,* March 13, 2012, p. 14.
97. Xie et al., "Addressing China's Water Scarcity," p. 42.
98. The 2030 Water Resources Group, *Charting Our Water Future.*
99. Ibid., p. 10.

4 FUTURE OF GAS MARKETS AND LNG TRADE IN ASIA

Less than a decade ago, the EIA forecast that U.S. demand for natural gas would increase at an average annual rate of 1.8 percent, from 22.7 trillion cubic feet (765 billion cubic meters) in 2001 to 34.9 tcf (988 bcm) in 2025.[100] With domestic production forecast to peak at 25.9 bcf in 2017 and then decline to 22.5 tcf in 2025, EIA projected in its base case that U.S. gas imports would nearly double from 3.7 tcf in 2001 to 7.8 tcf in 2025. Today that situation has reversed itself.

Less than six years later, U.S. shale gas production has massively and negatively impacted companies engaged in importing LNG into the United States. The effects are both first order—domestic production reducing the need for imported LNG—and now, second order—displacement of LNG contracted for the United States and even consideration of exporting of domestic gas as LNG. In 2010, the average capacity utilization for U.S. LNG terminals fell below 10 percent. The older terminals fared best, with the 1.035 bcf/d Everett, Massachusetts, plant achieving 40 percent use; Cove Point, Maryland, 12 percent; and Elba Island, Georgia, 17 percent. Elba Island added a 540 mmcf/d expansion in March 2010. Onshore LNG terminals at Golden Pass and Freeport, Texas, and Lake Charles, Cameron, and Sabine Pass, Louisiana, as well as offshore LNG operations serving New England and the Gulf all saw utilization at or below 5 percent.[101]

Facing plummeting imports (and revenues), several U.S. LNG receiving terminal operators have applied for permission to re-export contracted volumes. Just months after opening on a commercial basis, the Freeport LNG receiving terminal operators in August 2008 requested authorization from the U.S. Department of Energy to re-export up to 24 bcf of gas over a two-year period to Belgium, China, France, India, Italy, Japan, Korea, Spain, Taiwan, and the United Kingdom.[102] Sabine Pass also obtained permission to re-export LNG. In 2010, the Sabine Pass and Freeport LNG terminals re-exported 12 cargoes to Belgium, Spain, the United Kingdom, Brazil, Japan, and Korea for FOB prices of $5.22 to $7.70/MMBtu.[103]

As a second-order response, some LNG terminal operators and gas producers are considering taking what they see as undervalued domestic gas, liquefying it, and exporting the LNG to European and Asian markets where it could fetch two to three times the U.S. price. As early as August 2008, major U.S. shale gas producers began considering investment in U.S. LNG export facilities. While some companies are investing directly in export facilities, other shale gas producers prefer to get involved through long-term supply contracts.[104]

100. U.S. Energy Information Administration, *Annual Energy Outlook 2003,* Report no. DOE/EIA-0383(2003), released January 9, 2003, Washington, D.C.

101. Calculated from data provided by the U.S. Department of Energy, Office of Fossil Energy.

102. "Freeport LNG requests blanket authorization for exports," *Energy Current,* August 13, 2008.

103. Data provided by the U.S. Department of Energy, Office of Fossil Energy.

104. "Chesapeake Energy wants to export LNG," *Oil & Gas Journal,* October 14, 2010.

Since then a number of concrete export proposals have taken shape:[105]

- Sabine Pass LNG Terminal: In June 2010, Charif Souki, CEO of Cheniere Energy, owner of the Sabine Pass LNG terminal, announced that he was considering building liquefaction capability there.[106] The first phase of the export plant, which would include two trains with a capacity of about 1 bcf/d (10 bcm/y), would require state and federal approvals as well as signing satisfactory construction and long-term contracts with customers, Cheniere added. In August, Cheniere Energy Partners LP requested permission from the U.S. Department of Energy (DOE) to export up to 16 mmt/y for 30 years and from the U.S. Federal Energy Regulatory Commission to construct, in stages, four trains with a daily capacity of 2.6 bcf/d. The DOE approved the exports to any country with which the United States has entered into a Free Trade Agreement. Cheniere filed a separate application to export to countries that do not have a Free Trade Agreement with the United States.[107] Cheniere inked a deal with China's ENN Energy Trading Co. in November 2010 to export LNG from Sabine Pass to China starting in 2015,[108] and in February 2011 Japan's Sumitomo Corp. and Cheniere signed a memorandum of understanding under which Sumitomo could contract for bidirectional (import and export) capacity at the Sabine Pass terminal. Other Chinese and Spanish companies, as well as a U.S. financial institution, reportedly are negotiating with Cheniere for possible purchase of bidirectional capacity at Sabine Pass, which at 31 MMT/Y is the largest U.S. LNG terminal. In May 2011, the DOE approved Cheniere's separate application to export the gas to countries that do not have a Free Trade Agreement with the United States on a condition that licenses can be revoked any time for any reason, giving an ambiguous blessing to the industry.[109] In April 2012, the Federal Energy Regulatory Commission gave Cheniere its approval for the construction of an LNG export facility.
- Freeport LNG Terminal: In November 2010, Freeport LNG Development LP engaged Macquarie Group Ltd. to assist in building a $2 billion LNG export facility at its Freeport LNG terminal. The 1.4 bcf/d (10.5 MMT/Y) export operation is aiming for an early 2015 startup, which would allow Freeport to use the Panama Canal expansion to more easily service Asian markets.[110] Also looking at exporting by 2015 is Dominion Resources, Inc., operator of the Cove Point, Maryland, LNG terminal. Dominion would work with Norway's Statoil, which holds import capacity at Cove Point and production assets in the Marcellus shale.[111] Kitimat, British Columbia, shale gas production also could fuel LNG exports from Canada. An LNG export

105. Anna Driver, "Chesapeake studying LNG export facility investment," Reuters News Service, August 1, 2008, http://in.reuters.com/article/2008/08/01/chesapeake-call-idINN0146559320080801.

106. Moming Zhou and Paul Burkhardt, "Cheniere Plans to Export LNG from Sabine Terminal," Bloomberg.com, June 4, 2010, http://www.bloomberg.com/news/2010-06-04/cheniere-energy-may-build-natural-gas-export-plant-at-sabine-pass-terminal.html.

107. Isabel Ordonez, "Sabine Pass LNG Gets Energy Dept Approval to Export US Natural Gas," Dow Jones Newswires, September 9, 2010.

108. Alison Tudor, "China Bets Big on Gas Technology," *Wall Street Journal,* February 14, 2011.

109. U.S. Department of Energy Press Release, "Energy Department Approves Gulf Coast Exports of Liquefied Natural Gas," May 20, 2011, http://www.fossil.energy.gov/news/techlines/2011/11023-DOE_Approves_LNG_Export_Applicatio.html.

110. Ryan Dezember, "Macquarie, Freeport to Upgrade LNG Facility," *Wall Street Journal,* November 22, 2010, http://www.marketwatch.com/story/macquarie-freeport-plan-lng-upgrade-2010-11-22.

111. Edward McAllister, "Dominion eyes Cove Point LNG export by 2015," Reuters News Service, February 1, 2011, http://www.reuters.com/article/2011/02/01/lng-dominion-export-idUSN0122810220110201.

terminal already has been proposed at Kitimat, British Columbia, and The Calgary Herald has reported on rumors in the Canadian oil patch that two unnamed Canadian producers are considering a second west coast LNG export terminal.[112] Korea Gas Corp. (KOGAS) and Mitsubishi Corp. already had signed deals with Kitimat LNG Inc. in June 2009 for 1.8 MMT and 1.4 MMT, respectively, of Kitimat's eventual annual exports.[113] Houston-based EOG Resources, a partner with Apache in the Kitimat LNG project, also holds Horn River shale gas assets in British Columbia.

LNG export terminals in North America, as anywhere else, require significant up-front capital investments that generally can only be raised against long-term sales contracts with "take-or-pay" obligations. Also, any export of North American shale gas as LNG will have to compete with projects from strong existing LNG-exporting countries such as Qatar and Australia. Nonetheless, if Asian and European LNG prices continue at multiples of what North American producers can obtain domestically for shale gas production, there will be ample incentive to consider foreign markets. This incentive is magnified by the presence of Asian companies as investors in North American shale gas and as importers of LNG, who would welcome the portfolio and security diversity of one or more North American suppliers.

The greater impact of potential Chinese LNG imports likely will fall on prices rather than volumes. If China does succeed in developing significant shale gas production in the 2020–2030 period, and if its development reflects that in North America, the additional volumes will put downward pressure on natural gas prices and give China stronger bargaining leverage with natural gas exporters on price and other terms.

Meanwhile, the public debate is far from over as to what extent North American natural gas may alter the linkage between natural gas and oil prices for Asian markets.

The prospect for a robust LNG trade between North America and Asia may be heavily shaped by where an equilibrium will reside between the downward pressure on the LNG price level by successful shale gas development and the need for a price level that is high enough to entice and facilitate continued investments in shale exploration in the United States.

---

112. Deborah Yedlin, "Opinion: Encana deal sign of things to come," *Calgary Herald,* February 15, 2011, http://axilngas.com/Option__Encana_deal_sign.html.

113. Dan Healing, "LNG project signs Korean customer," *Calgary Herald,* June 2, 2009.

5 CONCLUSIONS

The pace of shale gas development in China, India, or elsewhere will likely depend on a combination of the following factors: (1) indigenous reservoir characteristics; (2) land use and mineral rights ownership; (3) expertise of local producers and the service sector; (4) commercial considerations, including infrastructure availability; (5) data availability and processing capability; and (6) regulatory environment, including applicable rules and laws governing water use.

Because of greater experiences in coalbed methane—although they are not directly transferrable for shale gas development—and infrastructure capacity, China would be farther ahead of India in developing its shale gas resources. China's primary concern is how to optimize the production and to achieve this goal on its own. The Chinese desire to develop the resources as indigenously as possible is leading the government to create internal competition among its national oil companies. Also, China has service companies that are already engaged in exploration and development (E&P) of unconventional gas resources. Consequently, the role of foreign companies in China would likely be limited to minority holdings in joint ventures with Chinese national oil companies.

The most significant driver for the Chinese and Indian shale gas industry would be the availability of technical expertise to manage and deploy advanced E&P technologies. We do not anticipate intervention by the U.S. government in controlling the outflow of such advanced E&P technologies. In fact, the technological advancement is more in such details as chemical additives in fluids. U.S. and western companies appear greatly aware of foreign interests in their E&P technologies and risks associated with a joint undertaking in countries that may not operate with the same contractual principles and intellectual property rights protections.

Equally key to the successful development is the availability of water. Shale gas E&P is a water-intensive process. Water is already a bottleneck for resource development in China and India. The ability to manage water resources and regulate waste water would hold a key to whether shale gas industry would take off successfully in China and India.

Shale gas in China and India, if developed successfully, will primarily meet domestic demand. Even their investments abroad, notably in North America, would benefit them by freeing up the LNG exports slated to North American markets. Further, Asian shale gas may put downward pressure on natural gas prices and give China stronger bargaining leverage with LNG exporters on price and other terms, narrowing the gap between North American and Asian natural gas prices. Asian shale gas development, therefore, eventually could affect global gas trade, primarily as LNG, but also development of transnational gas pipelines. The scale of this impact largely depends on the future trajectory of overall energy demand in China and India.

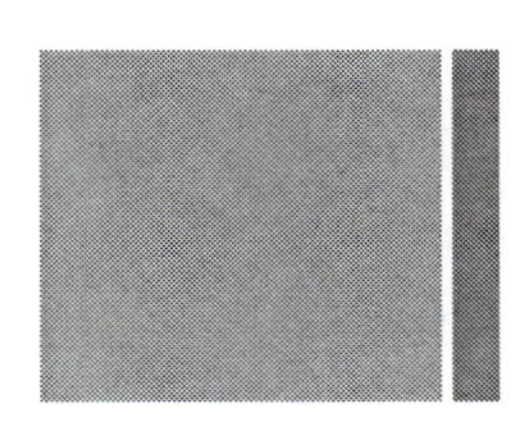

APPENDIX A
SHALE GAS TECHNOLOGIES

1. Horizontal drilling

Horizontal drilling is a technique often used in shale gas plays where the geology is significantly different from that of conventional oil and gas plays. While the costs of drilling horizontally can be two to three times higher than that of a vertical well, it allows for more cost-effective gas production because horizontal wells allow for greater contact with gas-bearing zones.[114] Shale gas operators increasingly rely on horizontal drilling to optimize recovery. For example, unlike traditional vertical-drilling techniques, horizontal drilling uses the same "drill pad" for multiple wells, and can extract more production from the well. It provides great access with a smaller footprint on the surface. Multiple horizontal wells from a single drilling pad could drain 200–640 acres, reducing disturbance to the natural habitat above and to the surrounding community.[115]

According to Oil & Gas Journal, FBR Capital Markets analyst Robert MacKenzie stated in a February 7, 2011, research note that, "Horizontally directed rigs currently account for 56 percent of the total U.S. land rig count, up from 17 percent in 2005 and 6 percent in 2000—34 percent of those are in shale plays (Barnett, Fayetteville, Woodford, Haynesville, Marcellus, Eagle Ford, and Williston Bakken)."[116] MacKenzie added that in 2011 companies would be seeking greater laterals, with Bakken operators pressing for more than 40 stages per well and lateral lengths in the Eagle Ford formations approaching 10,000 feet in some cases, up from 2010 when Eagle Ford had a lateral length of 6,000 feet and only 14 stages.[117]

Operational process and application. In the horizontal drilling technique, drilling begins with a central vertical wellbore descending to just above the shale play. Then, the drill makes a turn of approximately 90 degrees and drills horizontally.[118] This allows the wellbore to optimize contact area of the shale play. Several other technologies, adapted from other hydrocarbon exploration and development, support the advancement of horizontal drilling.

Rotary drilling rig and directional drilling offer another significant advantage in developing gas shales. Advances such as steerable down-hole drill motors that operate on the hydraulic pressure of the circulating drilling mud offer improved directional control. The newer tools are able to

---

114. Bettina Pierre-Gilles, "U.S. Shale Gas Brief," Phasis Consulting, September 2008, http://www.phasis.ca/files/pdf/Phasis_Shale_Gas_Study_Web.pdf.

115. U.S. Department of Energy, "Modern Shale Gas Development in the United States: A Primer," April 2009, http://www.rrc.state.tx.us/doeshale/Shale_Gas_Primer_2009.pdf.

116. Paula Dittrick, "Industry seeks new offshore rigs, longer onshore laterals in shale," *Oil & Gas Journal*, February 15, 2011.

117. Ibid.

118. Independent Oil & Gas Association of New York, "The Facts about Natural Gas Exploration of the Marcellus Shale," http://www.marcellusfacts.com/pdf/homegrownenergy.pdf.

drill directionally while rotating continuously, enabling a much more complex, and thus accurate, drilling trajectory. Continuous rotation and the improved navigation have also led to higher penetration rates and fewer incidents of the drill-string sticking.[119] The efficiency of the improved navigation shortens the time frame that the drilling rig is needed and helps lower costs overall.[120]

Casing and cement technology. Casing and cement are installed to protect freshwater aquifers from contamination and to prevent water leaking into the well.[121] As with conventional natural gas wells, the initial drilling goes through shallow groundwater into and below impermeable rock formations that separate the groundwater from the gas reservoir.[122] After completing the well to the target depth, the well is then cemented and a final casing put into place. Casing extends through the aquifer to prevent contamination. Concentric casing rings (steel and cement) provide additional buffers to protect groundwater supplies. Finally, a cement evaluation log is performed to measure the cement thickness, providing confirmation that the cement will prevent well fluids from bypassing outside the casing and infiltrating nearby aquifers.[123]

2. Hydraulic fracturing and fracking fluids

Hydraulic fracturing or "fracking" is a technology that helps to extract natural gas from reservoirs with low permeability and limited porosity.[124] Inducing hydraulic pressure in wells to fracture the formation and thus increase production by effectively increasing the wells' contact with the formation dates back to the late 1940s, but only with advances in the past three decades has it become feasible to use it for shale gas.[125] To make the unconventional gas more accessible, a mixture of water, sand or other proppants, and specially engineered chemicals, known as "fracking fluid," is pumped at high pressure into the natural gas reservoir, causing small fissures to form in the rock.[126] This process, also known as slick water fracking (as opposed to water-only fracking), allows the natural gas to flow out of the shale to the well in economically recoverable quantities. Proppants are used to keep the flow pathways open. Fracking is done multiple times in most gas plays, to release the gas from different areas and structures.

Water and sand constitute more than 98 percent of the fracking fluid, with the rest consisting of various chemical components.[127] The chemical/sand/water mixture is then pumped back out of the wellbore, leaving behind the proppants (typically sand or man-made ceramic), which keep the fractures open longer, which will permit more gas to be recovered. Once the mixture is removed,

---

119. Anthony Andrews et al., *Unconventional Gas Shales: Development, Technology, and Policy Issues,* Congressional Research Service, October 30, 2009, http://www.fas.org/sgp/crs/misc/R40894.pdf.

120. Robert Kennedy and Baker Hughes, "Shale Gas Challenges/Technologies over the Asset Life Cycle," presentation at the U.S.-China Oil and Gas Industry Forum, Washington, D.C., September 2010, p. 19–20.

121. U.S. Department of Energy, "Modern Shale Gas Development in the United States: A Primer."

122. Kelley Drye Client Advisory, "Hydraulic Fracturing: Short-Term Key Issues for Industry," December 10, 2010, http://www.kelleydrye.com/publications/client_advisories/0616.

123. Andrews et al., *Unconventional Gas Shales.*

124. Kelley Drye Client Advisory, "Hydraulic Fracturing."

125. Andrews et al., *Unconventional Gas Shales,* p. 17.

126. Ibid.

127. U.S. Department of Energy, "Modern Shale Gas Development in the United States: A Primer."

the pressure difference allows the gas to escape into the wellbore for recovery.[128] The pumped fluid, under pressures of up to 8,000 pounds per square inch, is enough to crack shale as much as 3,000 feet in each direction from the wellbore.[129]

Propane fracturing. In response to water quality and quantity issues associated with hydraulic fracturing (see section on "Emerging environmental issues"), some service companies have developed the use of propane or other liquid petroleum gases (LPGs) as an alternative fracturing agent. The process is essentially the same as hydraulic fracturing, but by using a prepared gel of liquid propane and sand or other proppants, propane fracturing does not require extensive use of local water resources. The advantage of using propane is that it allows for quicker production because the propane mixture can be extracted from the well faster than the water mixture used in hydraulic fracturing (1 to 2 days versus up to 5 days), and recovers nearly 100 percent of the fracking fluid versus 25–82 percent for hydraulic fracturing.[130]

3. Additional technologies

3D seismic. The application of high-quality three-dimensional (3D) seismic has been used in hydrocarbon exploration for some time, greatly improving the recovery rates. Within unconventional formations, 3D seismic is used to provide a detailed understanding of the formation above and below the reservoir.[131] This information has been key, as each formation is unique and requires a high level of knowledge of its particular geology.

---

128. Schlumberger, "Oilfield Glossary," 2012, http://www.glossary.oilfield.slb.com/Display.cfm?Term=permeability; Adam Orford, "Fractured: The Road to the New EPA 'Fracking' Study," *Marten Law*, September 17, 2010, http://www.martenlaw.com/newsletter/20100917-new-epa-fracking-study.

129. ClearOnMoney, "Unconventional Gas Background," January 20, 2010, http://www.clearonmoney.com/dw/doku.php?id=public:unconventional_gas_background.

130. Richhill, "Propane Fracking an Environmental Consideration," TransLoading.org, a Bulk Material Transfer Blog, July 14, 2010, http://transloading.org/propane-fracking-an-environmental-consideration/.

131. Kennedy and Hughes, "Shale Gas Challenges/Technologies over the Asset Life Cycle."

ABOUT THE AUTHORS

Jane Nakano is a fellow in the CSIS Energy and National Security Program. Her research interests include energy security and climate change in Asia, nuclear energy, shale gas development, and rare earth metals. Prior to joining CSIS in 2010, she was with the U.S. Department of Energy (DOE) and served as the lead staff on U.S. energy engagements with China and Japan. She was responsible for coordinating DOE engagements in the U.S.-China Strategic Economic Dialogue, U.S.-China Energy Policy Dialogue, and U.S.-Japan Energy Dialogue. She also worked on U.S. energy engagements with Indonesia, North Korea, and the Asia-Pacific Economic Cooperation. From 2001 to 2002, she served at the U.S. embassy in Tokyo as a special assistant to the energy attaché.

Nakano holds a bachelor's degree from Georgetown University's School of Foreign Service and a master's degree from Columbia University's School of International and Public Affairs. She is fluent in English and Japanese.

David Pumphrey is a senior fellow and deputy director of the CSIS Energy and National Security Program. He has extensive public-sector experience in international energy security issues and was most recently deputy assistant secretary for international energy cooperation at the Department of Energy. During his career with the federal government, he led the development and implementation of policy initiatives with individual countries and multilateral energy organizations. He was responsible for policy engagement with numerous key energy-producing and energy-consuming countries, including China, India, Canada, Mexico, Russia, Saudi Arabia, and the European Union. Pumphrey represented the U.S. government on the technical committees of the International Energy Agency (IEA) and the Energy Working Group of the Asia-Pacific Economic Cooperation Forum (APEC). He also represented the Department of Energy in negotiations on the energy-related sections of the U.S.-Canada Free Trade Agreement and the North American Free Trade Agreement.

Pumphrey received a bachelor's degree in economics from Duke University and a master's degree in economics from George Mason University. He has spoken extensively on international energy issues and testified before Congress on energy security issues related to China and India.

Robert Price Jr. is the president of International Risk Strategies, LLC, based in Tampa, Florida. He has advised clients on oil and gas exploration and development, refining, biofuels, power-generation plants, chemicals, and security and surveillance systems around the world.

Previously, Price served nearly two decades as a senior executive at the U.S. Department of Energy, most of that time as director of European and Asian affairs. Other positions held included acting deputy assistant secretary for national security policy, acting deputy assistant secretary for science and technology policy, and director of international science and technology cooperation. He represented the United States at senior meetings of the IEA, APEC, and the UN Economic Commission for Europe.

From 1981 to 1983, Price, as an IEA deputy division head, was responsible for *Natural Gas: Prospects to 2000* (1982), IEA's first natural gas publication, and for IEA's 1983 gas security study. He has authored several articles on international natural gas trade, participated as a speaker and panelist on international energy matters in the United States and abroad, and lectured at the College of Petroleum Studies, Oxford, and the U.S. Industrial College of the Armed Forces. He also has written articles on U.S.-Chinese cooperation in energy and the environment and was one of the authors of the IEA's *Developing China's Natural Gas Market: The Energy Policy Challenges* (2002). He has advised the governments of China and the Philippines on natural gas sector reform. He was an organizer and participant in the 2004 APEC LNG Workshop, which agreed and sent to APEC ministers recommendations of best practices to expand Asia-Pacific LNG trade.

A Vietnam-era veteran, Price served with U.S. Army Military Intelligence in Germany and worked as an Associated Press editor in Germany and as a reporter for the *Hartford Courant* newspaper. He is a member of the Academy of Political Science and the International Association of Energy Economics.

Molly A. Walton is a program coordinator and research associate with the Energy and National Security Program at the Center for Strategic and International Studies (CSIS). In this role she provides administrative support and research and analysis on a wide range of projects associated with domestic and global energy trends. Areas of interest include the energy-water nexus, unconventional fuels, environmental impacts of energy development, and geopolitical energy trends. Walton also serves on the editorial board of New Perspectives in Foreign Policy, a CSIS journal written by and for the enrichment of young professionals.

Prior to joining CSIS, Walton was a research analyst for Circle of Blue, an affiliate of the Pacific Institute, where she focused on the intersection of domestic water and energy issues. She previously interned at CSIS with both the Energy Program and the Global Strategy Institute. Walton received a master of arts in international relations and environmental policy from Boston University and holds a bachelor's degree in international relations and communications from Wheaton College (Illinois).